Oliver Teale

Teale's Light Line Phonography

The Easiest, Quickest and Most Legible System of Shorthand

Oliver Teale

Teale's Light Line Phonography
The Easiest, Quickest and Most Legible System of Shorthand

ISBN/EAN: 9783337270285

Printed in Europe, USA, Canada, Australia, Japan

Cover: Foto ©berggeist007 / pixelio.de

More available books at **www.hansebooks.com**

TEALE'S

→✳LIGHT LINE✳←

PHONOGRAPHY,

The easiest, quickest, and most legible

System of Shorthand.

———◆◆◆———

———◆◆◆———

Published at the **PHONETIC INSTITUTE.**

279 Smith St., Brooklyn, N. Y.

TEALE'S LIGHT LINE PHONOGRAPHY.

"Great praise is due to the gentlemen who report for the periodical press for the very accurate reports they furnish by means of their lengthy and defective systems of stenography." I believe this statement was first penned in the year 1845, and has been printed in Isaac Pitman's books ever since. It was aimed at the old-fashioned systems of shorthand. I now cast it back at its author, simply substituting the word "Phonography" for the word "Stenography." There is no doubt that some writers of old-fashioned phonography, after five or ten years' experience, succeed in doing excellent work with their lengthy and defective systems of shorthand, and when we consider that each outline in the Pitmanic systems* represents from six to fourteen different words, and that the unfortunate stenographer must guess which is the right one, the only wonder is that they can ever do good work.

In Teale's Light Line Phonography each outline represents one or two words, scarcely ever more than two. Of these two words one is generally a word of rare occurrence. For example: T written on the line represents the words It or Tea and nothing else.

In the Pitmanic systems, T on the line represents five words; T when shaded becomes D, and represents eight words more. As shading is almost impossible in rapid writing, T on the line in a Pitman reporter's notes represents thirteen different words, and the unfortunate reporter must guess which is the right one. In Teale's system, P on the line represents Hope or Pay and nothing else, while

---

* In speaking of the Pitmanic systems of phonography, we include all systems which are based on the Pitman alphabet, such as Munson, Graham, Longley, Marsh, etc.

in the Pitmanic systems P on the line represents six words; P when shaded becomes B, and represents eight words more, or fourteen in all.

Of course the student is told that he will know which is the correct word by the context, but, as almost every character in his notes is equally ambiguous, the context is a very unreliable guide, and the reporter is obliged to guess, and guess, and guess; and very often he guesses wrong. In Light Line Phonography guessing is out of the question—the student reads and does not guess.

Now about speed. A few minutes' examination of the engraved notes in this pamphlet will prove to the stenographer that Light Line Phonography is fundamentally the most rapid system of shorthand in the world.

In ordinary phonography there are thousands of contractions to be learned by heart; and if the student is not blessed with a remarkable memory he will surely fail in becoming an expert stenographer with those systems.

Light Line Phonography is fundamentally so rapid that very few contractions are necessary, and consequently there is very little to remember.

The remarkable legibility of Teale's Light Line Phonography makes it certain that in the near future this system of shorthand will be universally used. As a consequence of this remarkable legibility the old time practice of employers reading over and correcting letters written by stenographers will be entirely abandoned. An employer can put as much confidence in the accuracy of letters written in this system of shorthand as if they were dictated to a longhand writer, and consequently stenographers who use this system can command better salaries than any others.

O. TEALE,

PHONETIC INSTITUTE,

279 Smith Street, Brooklyn.

Consonants:–

B)   D /   P   G ——   H

J /   J   K ——   L   M

N   P \   R   S   T |

V (W   Y   Z   Ng

Th (Ch   Sh   R or V

Ex. 1.

The letters L, R, W, Y, and sometimes H,
are written upward, the remainder of the letters
are always written downward.

Vowels

- a as heard in the word ate
- E " " Eat
- I " " time
- ow " " how
- oz " " oil

— " — . . —

Ex. III

Ex. 6.

Day, aid, ache, gav, they, may, tame, lay, bay,
nay, pay, way, same, be, fee, key, knee, tea,
seed, reed, feed, need, meek, team time, tie,
die, my, write, ride, side, fight, bite, might,
cow, bough, thou, vow, out, couch, mouth, south,
boy, joy, toy, boil, soil, coil, foil.

This exercise and all others which are printed in ordinary type, should be transcribed,
several times in shorthand, the student repeat-
ing the name of each word as he writes it.

Additional vowels.

⌐	O as in hope	:	↓	oo as in food
⌐|	U as in up	:	∧|	U as in mute
⁄|	O as in not	:	⁄|	Aw as in all
⌐|	A as in mat	:	·|	E as in met
.|	I as in mit	:		

These vowels are written beside the letter
T, to show the difference in position thus:-

⌐ mat, ⌐ met, ⌐ mit, ⌐ not,

⌐ naught, ∨ foal, > fool,

⌐ nut, ⌐ newt.

The vowels a, e, i, are represented by dots
the only difference being that one is written
at the beginning of the consonant, another in
the middle and another at the end.

Ex. 7.

A consonant may generally be written half length to add T or D. thus: might, made, light, let.

Ex. 6.

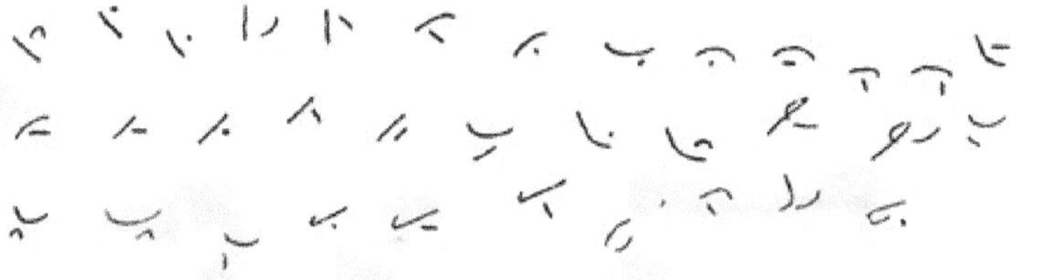

R may be added by writing the preceding consonant double length thus: bar far more power.

R may also be added by a small hook thus try, pray, draw.

The hook is used where R follows immediately after the preceding venng vowel.

[shorthand symbols]

A small hook on the same side as the R hook but at the end of the letter adds N:

A small hook on the opposite side to the N hook add L thus:

[shorthand symbols]

Ex 8.

[shorthand symbols]

Ex. IX.

Pan, pen, penny, pain, fan, Fanny, fun, funny,

dawn, sawn, lawn, Ben, Benny, bonny, **play**,

pale, plea, peel, clay, coal, fly, **file**,

pray, parp try, tire, dray, dare, **trim**,

term, train, turn, dry, dire.

A small hook at the end of a straight consonant, on the opposite side to the N hook, adds F or V thus:- ╱ deaf,

╲ puff, ⌐ cave.

S is written on the right hand or upper side of a straight letter. When S is written on the opposite side, it adds R at the beginning of a consonant, and N at the end, thus:- ╱ sat, ⌐ stray,

pies ╲ pines. ⌐

Ex. X.

Ex. XII.

Save, seed, said, stay, stray, stain,
sat, said, sod, cedar, sake , seek ,
stray, strive, straw, pains, pays, cough,
coffee, deaf, defence, dive, drive,
cave, waff, love, sigh, pine, pave,
proof, prove, dove, grieve, crave, skeen,
screen, positive, dine, dive, rave,

S may be written double length to add T,

ℓ   staff,   ⌐   last,   most,

S may be written trebble length to add Tr

◡   master,   ⌐   luster.

S may be written double size to addS or Z

O   size,   ⌡   faces,   ○   says,

A large hook at the end of a consonant

adds Shon,   ⌐   motion,   ⌡   fashion.

◡   nation,   ⌐   caution,

S may be added to the St loop, or to ss

thus:-   ⌐   lasts,   ⌐   successes.

A large hook at the beginning of a conson-

ant indicates an initial vowel thus:-

⌐   eat,   ⌐   ought,   ⌢   emotion,

⌡   omission,   ⌢   home,   ⌢   aim.

vowel
An initial may sometimes be indicated by
writing the first consonant vowel length.

(12)

Ex XII

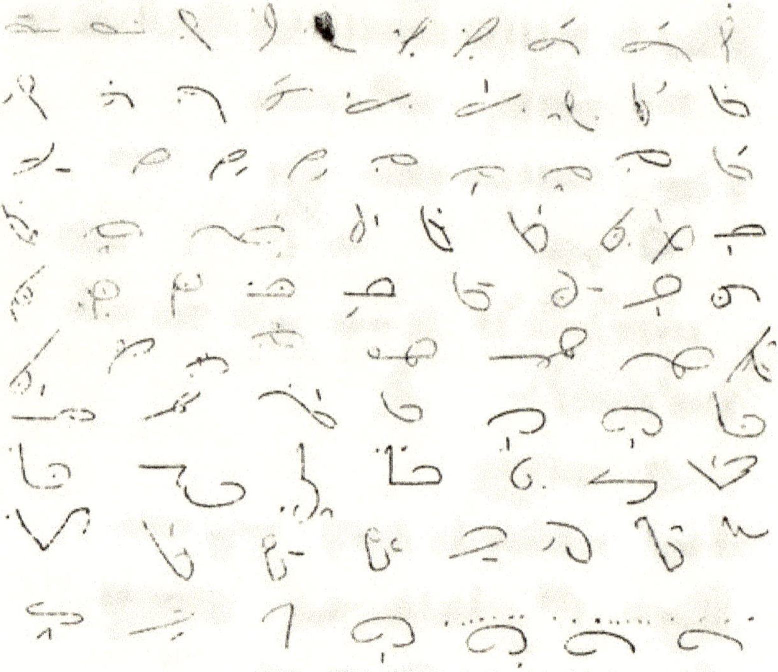

An initial vowel before s or a final vowel after s may be written by a small hook thus: ease, see, easy, fancy, essay,

(13)

I or E may sometimes be written before & thus :— ~ier cast
of Easter ○ cycs.

Ex. XIII

see, ease, easy, essay, use, sue, sow,
auspicious, aspire, aspiration, astonish,
astronomy, astronomer, ask, escape, ascend,
assent, easily, assumed, useful, esteem,
assign, assunder, assail, asleep, aslope,
eazel, isolate, isolation, astray, history,
user, easier, Easterly.

REPORTING.

In the last exercise dotted lines are
used, this is to show the difference in
position.   In reporting medial vowels are
generally omitted, and the consonant out-
lines written in three positions to indicate
the first vowell.

First position above the line, shows
that the first vowel is ah, aw, or i thus:-

) by, / had, / die, ⌒ lie.

Ex. XV.

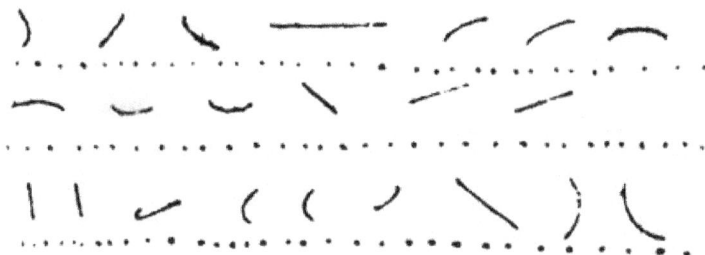

Ex. XVI.

By, buy, fie, guy, law, lie, my, am, nigh,
g aw, pie, paw, pa, raw, rye, tie, why,
thy, though, shy.

Second position on the line shows
the first vowel ise long, a long, e
short, a short thus:-) be, / day,

⎯ key, ⎯⎯ gay, ⌒ met,

Ex. xvl.

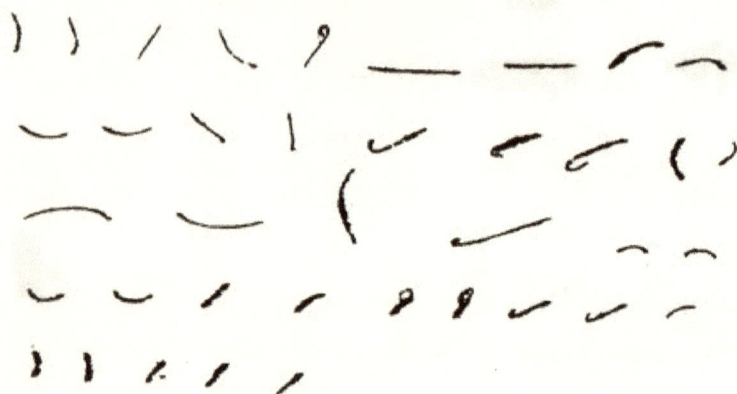

Ex. XVIII.

Bay, be, day, fee, hay, gay key, lay, may,

nay, knee. pay, way, yea, ye, there, where

met, meet, made, not, neat, need, debt, date,

dead, let. late, heat, hate, wet, wait, wit,

    Third position. through or under the line
shows that the first vowel is o, u, oi, oo,
or ow.

thus:- ..). boy,   ..)..obey,   /.. do,

Ex. XIX.

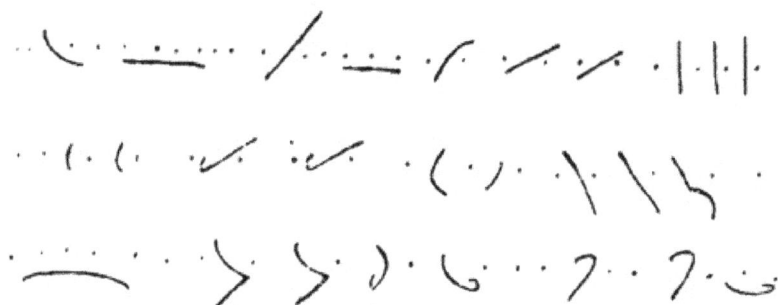

Ex. XX.

Boy, obey, do, due, few, go, joy, cow,

low, rue, row, to, too, two, view, vow,

woe, thou, show, poor, power, pure, more,

mode, food, feud, bun, fun, odd, ode,

owed, none, soon, sun, moon, noon, town,

W may be joined to a vowel thus:- ᴄ/

wed, ᴄ͡ will, ᴄ— wag, —ᴄ wake,

Y may be joined to a vowel, ⌣/ year

PHRASEOGRAPHY.

Words which frequently occur together in speech, may be joined in writing, by this means a considerable saving is effected.

A consonant may generally be written half length to add the.

After a W hook, or an initial vowel double hook, K may be written length to add R thus:- .... occur, ... accurate, occurrence, accuracy.

The V hook , and the initial vowel hook are never attached to the letter G.

M disjointed at the beginning of a word represents magni or magna thus, magnificence, magnanimous.

Word Signs and Contractions.

) 1 by, buy, above, 2 be, 3 but , obey, boy, 4 to be.

) 1 bought, bad, 2 ebbed, 3 body, 4 about.

) 1 better, bar, 2 bear, beer, 3 bore.

) 1 belong, 2 able, believe, belief, 3 obliged, oblige. 2___ obligation.

2, 1 absence, 2 baseness, business.

) 1 bank, bang, 2 being.

) 2 bring, 3 brother, number.

} barbarian, } barbarism.

) 3 beauty-iful,) beautifully behavior.) banker, -uptcy.

2 1 abundant, abandoned 2
L 2 between the, 3 obtained.

/ Had, die, 2 day, 3 do, due

7 3 dear, 3 during,

2 direct-ed, doctor

l different- ce, 6 differences,

difficult-y

advantage, danger

disadavntage, disadvantageous

disadvantageous

deliver-ed deliverence

distinct-ly-tion

distinguish-ed

distinctive

1 had not, 2 did not, do not

1 add-ed, aid-ed, 3 odd, ode

devolve-ed

discrete, 3 discord

half, if, 2 for, 3 from, few,

fact, 2 effect, 1 affect,

1 feature, 3 future,

February, 2 often, 3 Phonography

1 far, father, 2 tear, fare, 3 four

follow 1 fail, feel 2, full, fully 3

1 friend, 2 frequently,

1 find, 2 found, 3 fond

1 form, 2 free, 3 offer

affair, afire, afore, affure

familiar- ity

family, fortune-ate-ly

after, if the, 2 for the, 3 from

the, affair, afire, afore

formerly, formally

foundation, furnish,

———— 1 ago, 2 give, 3 go

—— 1 got, 2 get, 3 good

↩ 1 glad, ↩ great,

———→ 1 again, began, 2 gain, given, begin

3 gone, begun, gun

↩ 2 regulate-d, 3 glory,

↩→ 2 regulation, 3 glorification

——↙ govern-d-ment, ↙ governor,

J

/ 1 large, 2 joy, ⁊ age, hedge, 3

huge, ∫ adjoin, 2 agent,

/ 2 general-ly, ⅄ join, John, June

⌐ January, ↙ gentleman, 2 gentlemen

⤬ enlarge, ⁊ enjoy, ⁊ knowledge

⥾ agile, ∠ jollity, ⴭ journey

— 1 can, act, 2 come, 3 could, company

⌐ 1 quite, quiet, 2 question, 3 quote

↪ acquaintance, acquaint,

⇁ cannot, kind, 2 account,

⌐ car, 2 care, cure, cur, ⌐ court,

⟋ acquire-d, ⟋ awkward,

⌐ accordin- ly, 2 great, 3 court,

𝈍 catholic-ity, ⌐ carefully,

⟍ carpenter, ⌐ 1 cause, because

2 case , accuse,

σ— describe-d, 2 scripture, 3 secure

⌐ call, 2 equal-ly, 3 clue,

⌐ 1 caution, action, 3 occasion, auction

ʔ comparitive, ʔ comparitively,

⌐ commence, ⌐ commenced, ⌐ common,

⌐ command-ment, ⌐ comfort,

⟨ 1 all, law, lie, 2 will, lay.

3 low.

⟨ 1 allow, 2 ill, ail,

3 whole-ly, holy.

⟨ 1 while, ⟨ we will,

2 well, ⟨ with all,

⟨ allowed, allude,

3 held, hold, old.

⟨ 1 language, 2 English,

⟨ 1 all the, 2 let, late, little,

3 lot

⟨ 1 Lord, 2 letter, layer, 3 lower,

⟨ learn-ed, length,

⟨ 1 likely, 3 luckily,

⟨ 1 like, 2 lake, 3 look, luck,

⟨ 3 lowly, loyal,

⁀ 1 my, am, 2 him, may, 3 me, whom, home. ⁀ manufacture-d-ing

⌄ 1 myself, 2 himself, manufacture -ur-ing-ed, manufacturer ⁀

⌃ 1 met, made, meet, might, 3 mud, mode

⌃ 1 matter, mire, Mr., 3 mother,

⌁ 1 morning, 3 more-than,

⁀ 1 important-ce, improve-ment, ⋀

→ mistake, ⌁ impossible,

⌁ 1 mind, 2 may not, amount, 3 movement

⌁ 1 machine, ⌁ machinery,

⌁ 1 magnanimous-ly ⌁ magnificent-ce

→ mercy, ⌁ merciful;

⌁ miracle-ulous, ⌁ messenger,

⋀ misrepresent, ⌁ misrepresentation,

⌁ mortgage-d, ⌁ moral-ity,

⌁ mortality.

⌣ 1 In, 2 own, 3 on, know, no

⌣ only, ⌣ 1 not, night, 2 need,

end, 3 on the, 2 and the,

⌣ 1 in their, neither, 2 nor, near,

3 under, longer, on their, hundred

⌣ 1 another, 2 in our, inner, Henry,

3 honor-able, owner,

| entire, 2 enter-ed, ⌉ entertain,

⌮ 1 indefinite, 3 undefined,

⌡ 1 indifferent-ce, 2 endeavor

⌐ 1 inform-ed, 2 uniform-ed, 3 uniform-

ed, ⌐ 2 enable, ennoble, 3 unable

⌣ 1 interest, ⌣ 3 understood,

⌣ 3 understand-ing, ⌣ 3 uncertain

⌣uncerimonicus-ly, ⌣necessary,

⌣ United States, ⌣ unheard of,

(27)

1 part, 2 up, 3 upon, hope, 2 pay

happy, ↘ happiness,

happen, ↘ appoint, 3 point,

pound, ↘ appointment,

1 appear- ed, 2 pray, 3 principle-al,

1 appeared, pride, of our, 2 oppor-

tunity, 3 proud, ↘ applies,

applause, 2 please, 3 plus

1 private, approved, 2 profit, 3

proved, ↙ particular-ly,

pleasure, ↘ 3 provide-d,

painful, ↘ partnership,

party, par, 3 power, poor,

pure, 2 compare, ↘ 1 passion,

patient 2, ↘ patience,

peculiar-ity, ↘ pernicious,

1 are, or, 2 year, 3 our, hour

1 war, 2 your, 3 here, hear, her

1 art, heart, 2 heard, word,

ride, read (past tense) 2 read, 3 road, ⌒ rapid, 2 repeat

rather, 2 roar, rear, 5 run, !

Rome, room, ⌐ 1 Writer, 2 retire

recollect-ed, reflect,

recover-ed, regard-ed

require-d, relation, 3 lotion

remit-ed, remonstrate-d,

remonstrance, regret-ed,

record-ed, railroad,

railway, recommend-ed

refer, referrence

reform-ed, reformation,

(29)

○ 1 as, has, 2 is, his, 3 us

◡ as his, size, 2 is as, says, sees,

𝒪 1 as to, as to the, 2 first, 3 is to

is to the, Ϸ as it, 3 is it,

𝑔 consider-able, 𝘻 consideration,

𝑞 construct-ed, 𝘶 construction,

ρ satisfy- ied, sought, sight, 2 stay,

set, seat, sit, 3 suit,

℧ satisfaction, 2 station, 3 situation

𝜎 describe-d, 2 scripture, 3 secure,

𝜎⌐ description, ρ₃ society, 2 system

consist, ⟶ saying, ⌒𝑛 seeing,

3 sueing, ⌒◦ sequence

⟶ consequence, ⟶ consequent, second

⟲ secession, cessation, 𝒻 selfish,

⌒ forms the syllable self either at
.beginning or end of words.

\ at, 2 it, 3 out, | taught, tide, at the, 4 to the,

ᐯ 1 attach, at which, 2 teach, 3 touch.

ᚠ 1 at all, 2 till, tell, 3 until, to, all, ᒪ 1 at all events,

ᚷ 1 at their, 3 to their

ᒪ testimonial, ᒪ testimony,

ᐯ 3 tollerance, ᐯ tollerent,

ᒍ 1 at their own, torn, 3 turn,

ᒍ train, ᒌ 1 try, 2 truth, 3 true

ᚴ 3 two or three, ...outrun,

ᚦ 1 tried, 2 trade, toward,

ᒪ transcript, ᒪ transcribe,

ᒪ transcend, ᒪ transcendent,

ᒪ transgression, ᒪ transient,

ᒪ transfer, ᒪ transform-ed,

ᒪ transparent, ᒪ take the, take it,

(1 have, of the, 2 ever, 3 view, vow,

し 1 have been, 2 heaven, even,

ᒥ valuable, ꜱ available,

(evil, ꜱ various, varies,

3 over his, ꜱ 1 aver-red

3 however, hover-ed,

ꜱ 1 every, very, 3 over,

ꜱ 1 very much, 3 virtuous, over which,

3 virtually, ꜱ vision, evasion

ꜱ virgin, ꜱ Virginia,

ꜱ vegetable, ꜱ varnish-ed,

ꜱ volunteer, ꜱ voluntary-ly

ꜱ Why, away, 2 way, 3 one, woe,

ꜱ 1 Where, we are, 3 whether, weather,

ꜱ 1 want, 2 went, ꜱ 2 wind, wound

ꜱ wander, 2 winter, 3 wonder,

ꜱ 1 walk, 2 week, 3 work,

(1 tly, though, 2 they, them, 3 thou,
thee, without.

(2 there, they are, 3 other, through.

(1 either, author, 2 three.

1 these, 2 this, 3 thus, (those.

1 thank, 2 think.

1 authority, 3 throughout.

1 each, 2 which, 3 much.

2 chair, cheer, which are.

1 change,   2 chamber,

3 challenge,   1 charge.

1 child,   3 children.

3 church,   1 shall, 2 she,

wish,   3 show, issue.

1 short,   2 sure

2 usually,   2 surety.

(33)

ᴗ 2 thing, any, 3 long,   ⌒   1 beyond,

2 you, 3 young, ᴡ   2 owing,

⌐ 1 language, 2 English, ⱼ anxious

ᷠ anxiety, ⤳ younger, ⤴ youngster

Vowels.

ᐟ 1 I, of, 2 the, 3 before,

⌐ 1 that, 2 should, he, 3 who, how,

ᐠ 1 was, 2 owe, Oh, 3 out,

ᐸ 1 we, with, 2 when, 3 were,

ᗡ 1 what, 2 would, ᐸ   ultimate-ly

ᐸ ultimo, ᐸᷜ unanimous

ᐸᷜ anonymous, ᷑ uncommon,

ᑉ illegible, ᐸᷜ illegal,

ᑉ enemy, ᷜᐯ immortal-ity

ᷜ immigration, ᷜᷡ emigration,

ᷱ 1 a, an, and, 2 say, 3 so,

Try

The Wolf and the Lamb

TRY.

There was once a good little dwarf
named **Try**, who was so powerful that he
overcame everything that he attempted, and
yet was so small that people laughed when
they were told of his wonderous powers.
But the tiny man was so kind at heart
and loved so much to serve those who
were less able than himself, that he would
go and beg of those who knew him better to
plead for him that he might be allowed to
help them out of their troubles, and
when once he had made them happy, by
his noble deeds they no longer despised
him, or drove him away but loved him as
their best friend; yet the only return

this good dwarf sought for all his services

was that when they knew any one who wanted
a helping hand , they would say a good
word in his favor and commend them to Try.

The Wolf and the Lamb

One hot day a wolf came to quench his
thirst at a clear brook that ran down the
side of a hill. By chance a young lamb
stood there. The wolf had a desire to eat
her, but felt some qualms so for a plea he
made out that the lamb was his foe.
" Stand off from the banks sir said he,
for as you tread them you stir mud in the
stream, and all I acn get to drink is thick
and foul. The young lamb said in a mild tone
that she did not see how that could be the
case , as the brook ran down to her , from
the spot where stood. Oh said the wolf
how dare you drink of it at all till I have
had my fill? Then the poor lamb told him that
as yet her dams milk was both food and drink
to her. Be that as it may said the wolf
you are a bad lamb for last year I heard
that you spoke ill of me and all my race.
Last year dread sir, quoth the lamb, why
I have not yet been shorn, and at the time
you name I was not born.

REMINGTON STANDARD
⚡TYPE-WRITER⚡

OVER 40,000 IN DAILY USE.

WYCKOFF, SEAMANS & BENEDICT,
327 BROADWAY, N. Y.

Speed Contests, Cincinnati, July 25, '88. Highest speed on legal work. New York Aug. 2 '88 Highest, speed on correspondence. Toronto, Aug. 13, '88. (Interna'l Tournament for World's Championship.) 1st and 2nd Prizes, business correspondence. 1st and 2nd Prizes, legal testimony, gold and silver medals. Send for circular and price-list.

TEALE'S LIGHT LINE PHONOGRAPHY.

On account of its wonderful legibility and speed can be learned in half the time required by ordinary shorthand.

Terms. $15. for three month's course.   Type-writing $10.

Manuel or Reader, post-paid 60c. Text Book, complete, $1.

O. TEALE, PHONETIC INSTITUTE, 279 Smith Street, Brooklyn.

New York Office. 63 Broadway.

"Underwood's" Superior Type-writer Ribbons,

Caligraph Ribbons, Carbon Paper.

J. UNDERWOOD & CO.,

30 Vesey St. New York.   10 Johnston St. Toronto.   165 LaSalle St. Chicago.

"IT STANDS AT THE HEAD."

THE NO. 2 CALIGRAPH,

Is the only writing machine of its class which produces each character at a *single stroke*, doing away with the old-fashioned '*shift keys*" and "*false motions*" and it is becoming immensely popular for its *Speed. Simplicity and Manifolding Power,*

THE AMERICAN WRITING MACHINE COMPANY,

HARTFORD, CONN.

New York Office, 237 Broadway.

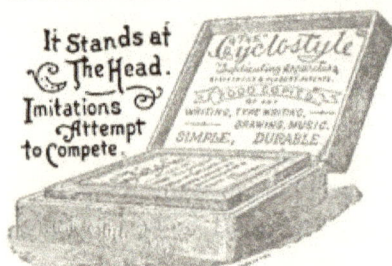

Light Line Phonography is one of the greatest inventions of the age, it is almost as legible as long-hand Writing, and yet is one of the quickest systems of Shorthand in the world.

It can be learned in half the time required by ordinary shorthand.

Private Lessons, day or evening, 63 Broadway, New York, and 279 Smith St., Brooklyn.

Manuel or Reader, paper covers 60c, cloth 75c.

Text Book, complete. $1,00.